I DON'T BELIEVE IN ASTEROIDS

A
HURRY UP PLEASE IT'S TIME
COLLECTION

This book is for Alex and Asher
and for all the other children
who will reap what we sow.

Copyright © 2015 by Kevin Kite & Michelle McCauley

Versions of the comics in this book were originally published on the website www.hurryuppleaseitstime.com and are copyright © 2014 and 2015 by Kevin Kite and Michelle McCauley.

All rights reserved.
Published by Inkfall Studios, New Haven, Vermont.

ISBN: 978-0-9965849-9-9

First Edition, July 2015
Book Design by Kevin Kite
Printed in the United States

CONTENTS

Chapter 1
There's somethin' I gotta do in the woods.......1

Chapter 2
The return of the living 47

Chapter 3
Why did you burn this one? 73

Chapter 4
It's your time, that's all! 95

Chapter 5
Get big or get out 111

Chapter 6
Your things will be safe in my pockets 127

Chapter 7
Clear cut your wood lot...................................157

Chapter 8
So, get up..171

Chapter One:

There's somethin' I gotta do in the woods

There's somethin' I gotta do in the woods

CLOSER

PROCRASTINATION

There's somethin' I gotta do in the woods

There's somethin' I gotta do in the woods

URBAN LIVING

URBAN LIVING AFTER THE RISING

There's somethin' I gotta do in the woods

TREE

THE BIOENGINEERS WON THE ENVIRONMENTALISTS OVER WHEN THEY PRODUCED THE SOLAR PANEL WIFI TREE

SLOWDOWN

SENATE

There's somethin' I gotta do in the woods

MITES

There's somethin' I gotta do in the woods

VALUE

THERE'S SOMETHIN' I GOTTA DO IN THE WOODS

REMINDER

There's somethin' I gotta do in the woods

METHANE

There's Somethin' I Gotta Do in the Woods

There's somethin' I gotta do in the woods

BURGERS

EMBEDDED

PUBLIC SERVICE ANNOUNCEMENT
LET'S CLEAR THIS UP.

There's somethin' I gotta do in the woods

REHASH

There's somethin' I gotta do in the woods

There's somethin' I gotta do in the woods

SENSE

There's somethin' I gotta do in the woods

THAT MAKES SENSE, DOESN'T IT? [5]

FOOTNOTES AND FINE PRINT:

[1] On January 21, 2015, the Senate passed an amendment to the Keystone XL Pipeline Act by a vote of 98 to 1, stating, "it is the sense of Congress that--
(1) climate change is real; and
(2) human activity contributes to climate change."

[2] The same day, the Senate failed to pass an amendment stating "it is the sense of Congress that--
(1) climate change is real; and
(2) human activity contributes to climate change." (Sixty votes are required to move legislation in the Senate, and the amendment failed with a final vote tally of 59-40.)

[3] Another amendment stating "it is the sense of Congress that--
(1) climate change is real; and (2) human activity *significantly* contributes to climate change" failed to pass by a vote of 50-49.

[4] On January 22, the Senate voted 56-42 to table an amendment which would have said:
(1) climate change is real;
(2) climate change is caused by human activities;
(3) climate change has already caused devastating problems in the United States and around the world;
(4) a brief window of opportunity exists before the United States and the entire planet suffer irreparable harm; and
(5) it is imperative that the United States transform its energy system away from fossil fuels and toward energy efficiency and sustainable energy as rapidly as possible.

[5] No, not really.

SAM

There's somethin' I gotta do in the woods

SENATE

THERE'S SOMETHIN' I GOTTA DO IN THE WOODS

AGREE

IN THE GARDEN

INACTION

STRANGELY, EXPENSIVE IDEAS REQUIRING ACTION...

...USUALLY TRUMP INEXPENSIVE IDEAS REQUIRING INACTION.

There's somethin' I gotta do in the woods

RESEARCH VESSEL

WELCOME ABOARD R/V *EARTH*, WHERE YOUR PASSAGE IS MANDATORY AND YOUR FARE IS NON-REFUNDABLE.

ANYTHING

There's somethin' I gotta do in the woods

OUTREACH

Sadly, Drs. Doom, Gloom, and Soon were rarely invited in.

There's Somethin' I Gotta Do in the Woods

BEAR

I'M OUTTA HERE! IT'S TOO DARN HOT, AND THERE'S SOMETHIN' I GOTTA DO IN THE WOODS!

one day, bear had simply had enough.

FORECAST

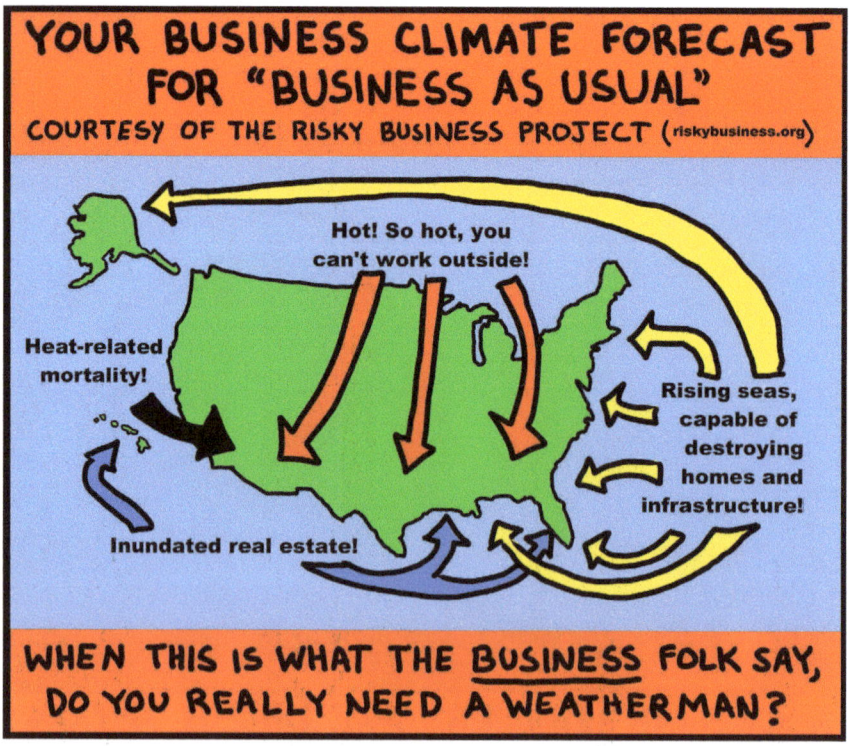

SHOWER

There's somethin' I gotta do in the woods

SCIENTISTS

TARDIGRADE

There's somethin' I gotta do in the woods

STRANDED ASSET

OPTIMIST/PESSIMIST QUIZ

There's somethin' I gotta do in the woods

LAST

There's somethin' I gotta do in the woods

Chapter Two:

The Return of the Living

SIGNPOST

"I'VE GOT AN UNEASY FEELING ABOUT THIS..."

CHOCOLATE

BIF HAD ALWAYS LOVED CHOCOLATE ICE CREAM UNTIL HIS FATHER'S FRIEND FROM THE FACTORY SAID THE CHOCOLATE WAS MADE FROM VANILLA GONE AWRY.

THE RETURN OF THE LIVING

SPLICE OF LIFE

HAROLD'S SPLICING OF LEECH AND PYTHON PROVED PARTICULARLY PROBLEMATIC

REGRETS

THE RETURN OF THE LIVING

RETURN

FOR LONG AFTER, THEY NERVOUSLY FEARED THE RETURN OF THE LIVING.

THE RETURN OF THE LIVING

FLUNKED

NARRATIVE

CROSSING BOUNDARIES

THE RETURN OF THE LIVING

SKEPTICAL

THE RETURN OF THE LIVING

GROWING

DESPITE HER BEST EFFORTS, HENRIETTA WAS UNABLE TO RESOLVE HER GROWING INSECT PROBLEM.

THE RETURN OF THE LIVING

UNCERTAIN

EVEN WITH MUTANT GENES TO GUIDE THEM

THE EARLY SEXUAL REPRODUCERS WERE UNCERTAIN, AT FIRST.

TRANSISTOR CELL PHONE

HORACE

OPEN

THE RETURN OF THE LIVING

HUMIDITY

THE RETURN OF THE LIVING

IDEA PICTURE

THE RETURN OF THE LIVING

TEA HERE NOW

THE RETURN OF THE LIVING DONUTS

THE RETURN OF THE LIVING BILLIONAIRE

ADVERSE IMPACT

THE RETURN OF THE LIVING STARFISH

THE RETURN OF THE LIVING

THE RETURN OF THE LIVING

THE RETURN OF THE LIVING

Chapter Three: Why did you burn this one?

WHY DID YOU BURN THIS ONE?

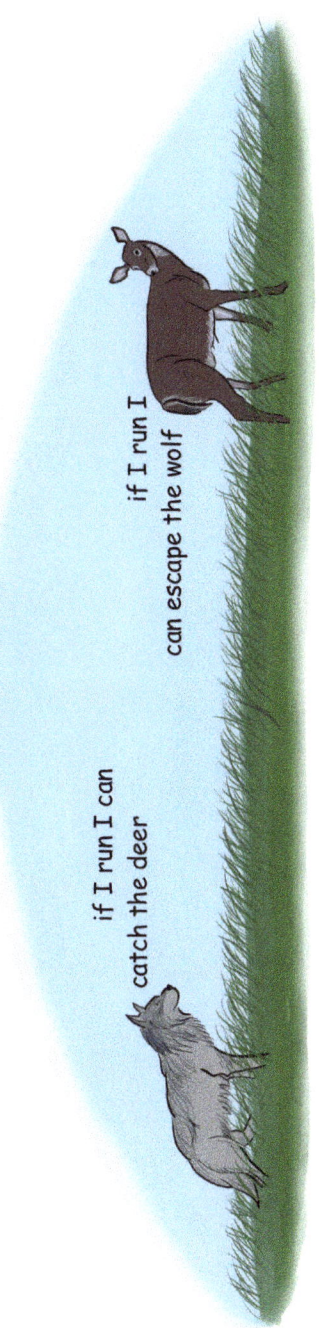

RUN

if I run I can catch the deer

if I run I can escape the wolf

just run.

Why did you burn this one?

WHY DID YOU BURN THIS ONE?

PROGRESS

75

Why did you burn this one?

WHY DID YOU BURN THIS ONE?

EYES

AS HOWARD HURRIED HOME, HE FELT ALL EYES ON HIM.

Why did you burn this one?

Why did you burn this one?

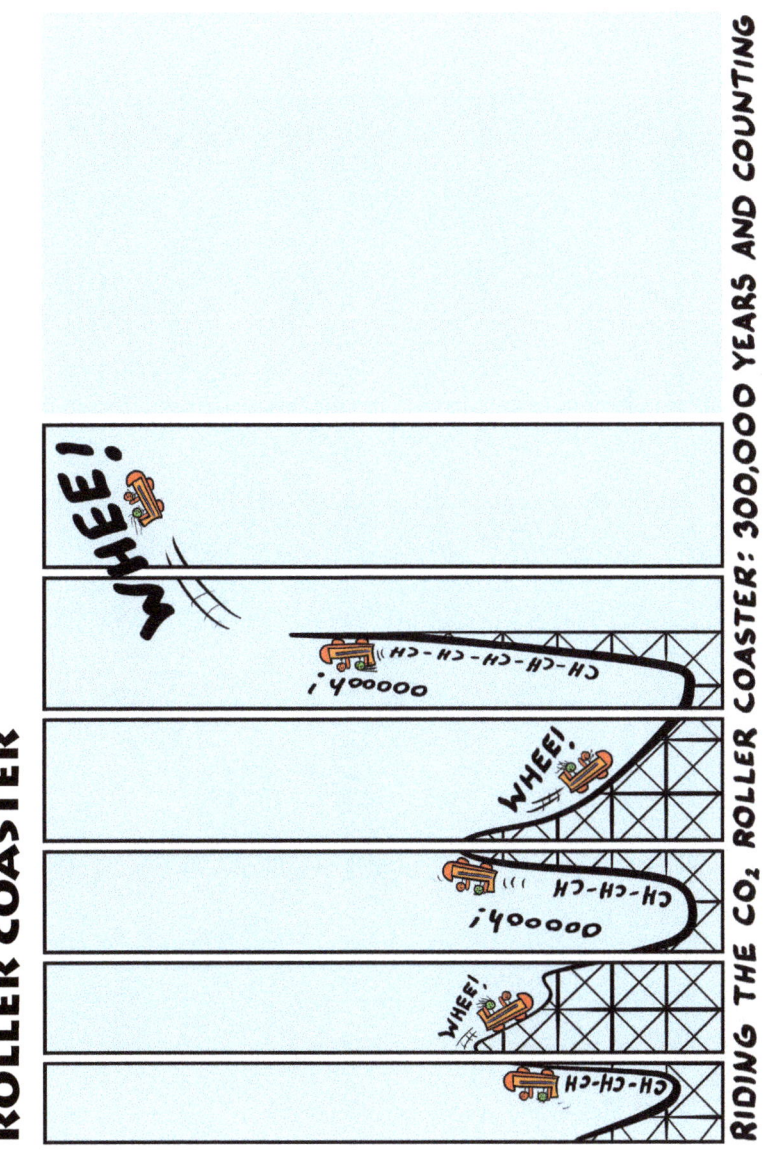

WHY DID YOU BURN THIS ONE?

LUNCH

WILBUR TOOK EXTRA PRECAUTIONS AT LUNCH AFTER HE LEARNED COCA-COLA HAD REMOVED THE FLAME RETARDANTS FROM ITS DRINKS.

WHY DID YOU BURN THIS ONE?

Why did you burn this one?

DEFENSE

Despite the determined onslaught of the saxophone and the trumpet, the flutes and the guitar mounted a spirited defense

WHY DID YOU BURN THIS ONE?

Why did you burn this one?

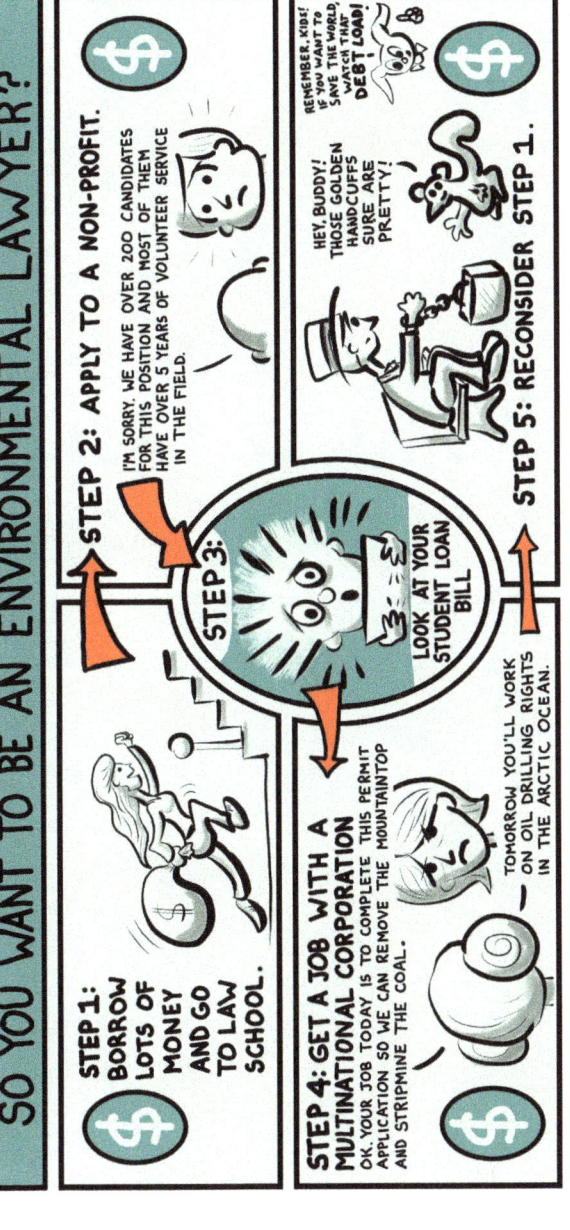

Why did you burn this one?

WHY DID YOU BURN THIS ONE?

CARB COUNTER

WHY DID YOU BURN THIS ONE?

WHY DID YOU BURN THIS ONE?

WHY DID YOU BURN THIS ONE?

Why did you burn this one?

Why did you burn this one?

FIRE CODE

Chapter Four:

It's your time, that's all!

RESOLUTIONS

BIRTHDAY

It's Your Time, That's All!

INTO YOU

GLOBAL DIVESTMENT DAY

IT'S YOUR TIME, THAT'S ALL!

JOY

PROGRESS ROBS SALLY OF ONE MORE CHILDHOOD JOY

A DAY OF ONE'S OWN

IT'S YOUR TIME, THAT'S ALL!

IN MEMORY . . .

LABOR DAY

PRE-HALLOWEEN JITTERS

IT'S YOUR TIME, THAT'S ALL!

VAMPIRE

IT'S YOUR TIME, THAT'S ALL!

TRICK

HALLOWEEN GETS SCARIER ALL THE TIME

IT'S YOUR TIME, THAT'S ALL!

PIE

TURKEY THANKSGIVING

It's your time, that's all!

SPENT

CONUNDRUM

LOCATION, LOCATION, LOCATION

Chapter Five:

Get big or get out

REBOOT

BALANCE

TEND

DANGEROUS MEMES OF OUR TIME 3:
IT'S BEST TO TEND ONE'S OWN...

...GARDEN

CHINA

EXPECTATION DAMAGES

DATA

CREDIBLE

WISHFUL THINKING

Get big or get out

SELF-RELIANCE

Day 9,788

NECESSITY

SORRY, BOYS, SIZE MATTERS

GROWTH

Get big or get out

Chapter Six:

Your things will be safe in my pockets

BABY

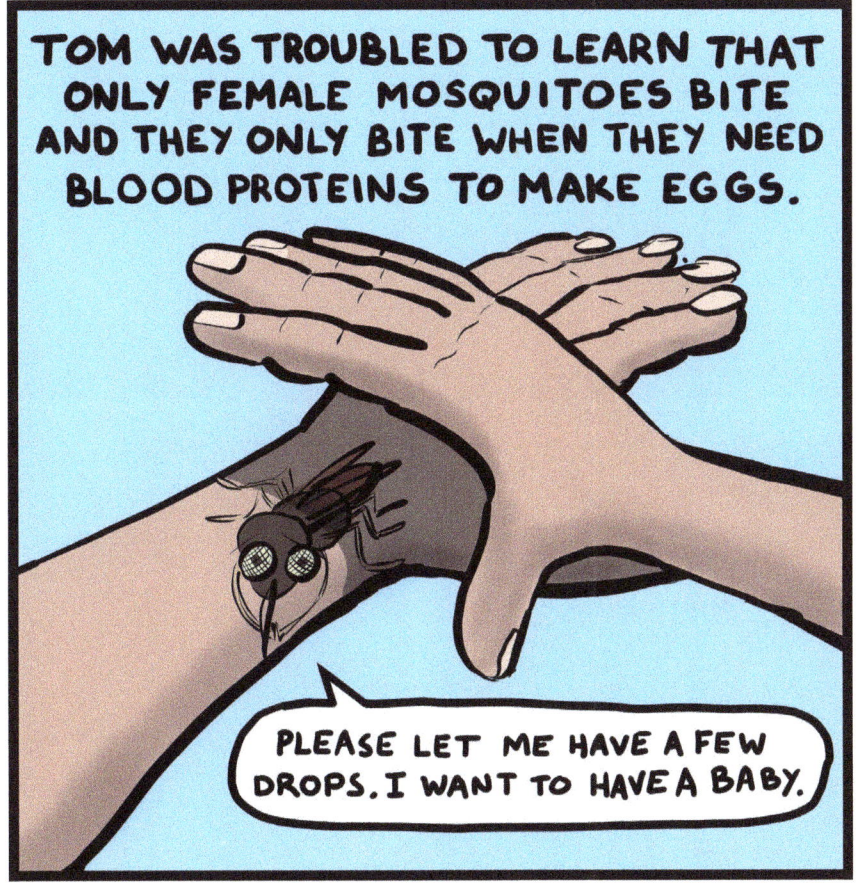

SIMPLIFY

Your things will be safe in my pockets

RATIONAL ACTOR

YOUR THINGS WILL BE SAFE IN MY POCKETS

BIOENGINEERS

BELIEVE

Your things will be safe in my pockets

The Supreme Court gave the corporation a voice in *Citizen's United*; now, in *Hobby Lobby*, they give it a **soul!**

YOUR THINGS WILL BE SAFE IN MY POCKETS

INVISIBLE HAND

Your things will be safe in my pockets

Your things will be safe in my pockets

WAX

ROGER LIKED WAX BOTTLES UNTIL HE TRIED TO MAKE ONE AT HOME.

CUBE

HOBBLING

YOUR THINGS WILL BE SAFE IN MY POCKETS

DR. PEABODY

FULL

PROGRESS

YOUR THINGS WILL BE SAFE IN MY POCKETS

PHENOTYPES

THE PHENOTYPES OF IN-LAWS' OFFSPRING CAN NONETHELESS PROVE ATTRACTIVE TO THE OFFSPRINGS' MATES.

NURTURE

RACKET

SOAR

ACID

YOUR THINGS WILL BE SAFE IN MY POCKETS

EXPELLED

NEURON

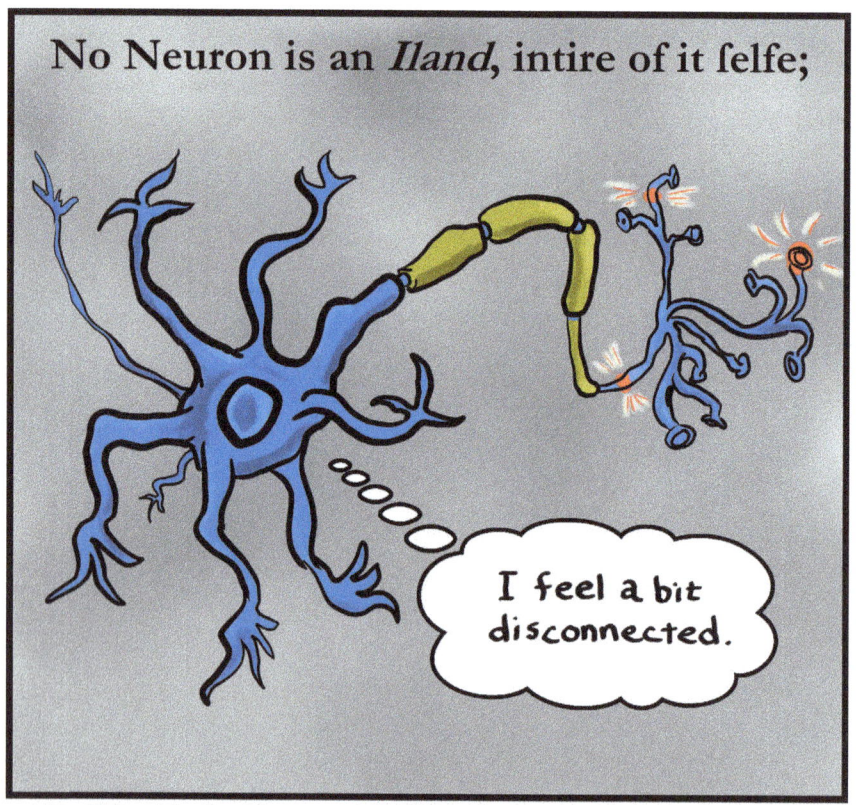

IF JOHN DONNE HAD BEEN A NEUROSCIENTIST.

EXHIBIT

PEDESTRIAN

DR. HUGO MAKES AN UNFORTUNATE ASSERTION AT THE TRANSPORTATION POLICY CONFERENCE.

YOUR THINGS WILL BE SAFE IN MY POCKETS

YES, BUT

PROMISING

SEO

SEARCH ENGINE OPTIMIZATION PARENTING

TALK

WHILE ARCHIBALD WAS IN THE KITCHEN MAKING TEA, VERONICA WAS IN THE DEN MAKING MISCHIEF

APPLAUD

CHILDHOOD TRAUMA NO. 37: MY MOTHER NEVER APPLAUDED AT MY RECITALS

Chapter Seven:

Clear cut your wood lot

EVERYTHING

STUMPED

Clear cut your wood lot

Clear cut your wood lot

SURF

OFFSET

Clear cut your wood lot

PAY ME

Clear cut your wood lot

CLEAR CUT YOUR WOOD LOT

CHAIN SAW

Clear cut your wood lot

CUSTOM

Clear cut your wood lot

SHARING

Clear cut your wood lot

FACTS

169

Chapter Eight:

So, get up!

SO, GET UP

BED

SO, GET UP

SON — ALL WE HAVE IN THIS WORLD IS THE ARBITRARY IMPOSITION OF FORM ON AN OTHERWISE MEANINGLESS VOID, WHICH TEMPORARILY LENDS PURPOSE TO OUR ACTIONS!

SO, GET UP

"IF YOU DON'T GET UP, THE TRUE POINTLESSNESS OF ALL ACTION WILL BECOME UNBEARABLY APPARENT..."

SO, GET UP

SO, GET UP

SO, GET UP

PERSPECTIVE

SO, GET UP

SO, GET UP

Notes

The answer to the question, "Where do you get your ideas?" is: all around us. Most of the comics in this book were made in response to something we saw, read, heard about, or experienced.

Usually, the event gets filtered through our brains and Kevin's pen. But sometimes, you don't really have to do much more than repeat the story.

"direct action" (page 37) was made after we read about a herd of Norwegian Deer who occupied the Stallogargo Tunnel to take advantage of the cooler temperatures inside the tunnel, forcing the human population to take a detour. http://www.thelocal.no/20140806/shade-seeking-reindeers-shut-highway-tunnel.

"skeptical" (page 55) summarizes a methodology scientists used to study the flight patterns of geese. *See* C.M. Bishop et al. The roller coaster flight strategy of bar-headed geese conserves energy during Himalayan migrations. **Science**. Vol. 347, January 16, 2015, p.250. doi: 10.1126/science.1258732.

Finally, we want to note that "favorite" (page 93) came from an idea by Asher Kite. Asher also drew "idea picture" (page 62) and wrote and drew "president" (page 64).

Read more comics at
www.hurryuppleaseitstime.com

About the Authors

Kevin Kite has been an artist since he could hold a pencil (his earlier crayon work is lost to us given the humid climate of his native Florida). He spent a few years detouring into other areas (don't ask), but now he draws, writes, takes photos, and stares through windows in Vermont, where everything looks a bit sillier every day.

Michelle McCauley does not draw anything—ever. She does, however, teach conservation psychology at Middlebury College. Her current hobbies are offsetting her own carbon use one tree at a time in her backyard (she has a really big backyard) and hanging clothes out to dry on the line, in a field, in the dark.

INDEX OF COMICS BY TITLE

acid 146	exhibit 149
a day of one's own 100	expectation damages 115
adverse impact 67	expelled 147
agree 30	eyes 77
agreement 24	facts 169
anything 34	favorite 93
applaud 155	fire code 92
asteroids 78	flunked 52
baby 127	forecast 38
balance 112	full 140
bear 36	fun physics 80
bed 171	glacial 88
believe 132	global divestment day 98
billionaire 66	growing 56
bioengineers 131	growth 124
birthday 96	history 76
burgers 18	hobbling 138
carb counter 87	homeland security 82
chain saw 164	horace 59
china 114	humidity 61
chocolate 48	idea picture 62
closer 1	inaction 32
conundrum 108	in memory 101
credible 117	in the garden 31
crossing boundaries 54	into you 97
cube 137	invisible hand 134
custom 166	joy 99
data 116	labor day 102
day 9,788 120	last 44
defense 83	location, location, location 109
direct action 37	lunch 81
donuts 65	march 91
dr. peabody 139	methane 14
earthworm 90	mites 10
embedded 19	narrative 53
environmental lawyer 85	necessity 122
everything 157	neuron 148

nurture	143
offset	161
open	60
optimist/pessimist quiz	43
our evolving energy crisis	86
outreach	35
pay me	162
pedestrian	150
perspective	179
phenotypes	142
pie	106
pre-halloween jitters	103
president	64
prey	74
procrastination	2
progress	75
progress	141
promising	152
public service announcement	20
racket	144
rational actor	130
reboot	111
regrets	50
rehash	22
reminder	12
research vessel	33
resolutions	95
return	51
roller coaster	79
run	73
sam	28
scary	84
scientists	40
self-reliance	119
senate	9
senate	29
sense	26
seo	153
sharing	168
shower	39
signpost	47
simplify	128
skeptical	55
slowdown	8
soar	145
sorry, boys, size matters	123
spent	107
splice of life	49
starfish	68
stargazing	89
stranded asset	42
stumped	158
surf	160
sweden	135
talk	154
tardigrade	41
tea here now	63
tend	113
transistor cell phone	58
tree	7
trick	105
turkey thanksgiving	106
uncertain	57
urban living	6
value	11
vampire	104
vanishing pants	4
wassup	16
wax	136
wishful thinking	118
yes, but	151

www.ingramcontent.com/pod-product-compliance
Lightning Source LLC
Chambersburg PA
CBHW062058290426
44110CB00022B/2634